Like Ants on the Bottom of the Abyss

A CHRISTIAN EARTH SCIENTIST EXPLAINS
THE CONCEPT OF CLIMATE VARIABILITY
WITH AN ECCENTRIC PERSPECTIVE

DRUKELL B. TRAHAN

WestBow Press
A DIVISION OF THOMAS NELSON
& ZONDERVAN

Copyright © 2017 Drukell B. Trahan.

All rights reserved. No part of this book may be used or reproduced by any means, graphic, electronic, or mechanical, including photocopying, recording, taping or by any information storage retrieval system without the written permission of the author except in the case of brief quotations embodied in critical articles and reviews.

WestBow Press books may be ordered through booksellers or by contacting:

WestBow Press
A Division of Thomas Nelson & Zondervan
1663 Liberty Drive
Bloomington, IN 47403
www.westbowpress.com
1 (866) 928-1240

Because of the dynamic nature of the Internet, any web addresses or links contained in this book may have changed since publication and may no longer be valid. The views expressed in this work are solely those of the author and do not necessarily reflect the views of the publisher, and the publisher hereby disclaims any responsibility for them.

Any people depicted in stock imagery provided by Thinkstock are models, and such images are being used for illustrative purposes only.
Certain stock imagery © Thinkstock.

ISBN: 978-1-5127-7967-7 (sc)
ISBN: 978-1-5127-7968-4 (hc)
ISBN: 978-1-5127-7966-0 (e)

Library of Congress Control Number: 2017904150

Print information available on the last page.

WestBow Press rev. date: 3/27/2017

Contents

Acknowledgements vii
Introduction ix

Miracle Planet 1
Universe 5
Randomness 11
Change 15
Measurement 19
Scale 23
People 27
God 31
Bandwagon 35
Man-Made 41
Forecasting 45
History 49
Historical 53
Climate Change 101 57

The Data	61
Example	65
Jet Streams	67
Sea Level	71
Extinction	75
Footprints	79
The Gases	81
Matrimony	85
References	89
Appendix	93

Acknowledgements

First of all, I am thankful for the spirit that led me to write this manuscript and get it out there for any criticism or praise it may receive. I hope the spirit came from a good source of inspiration and drive. I am thankful to my wife for putting up with whatever inconveniences I cause due to my thought processes, which never end. I am thankful for my family, grandchildren, in-laws, friends, educational associates, and occupational associates for shaping me through all the inspirations I've encountered through life.

In particular, among these are my brothers and sisters from Pi Kappa Alpha fraternity Zeta Psi chapter, my brothers from my once-a-month spiritual dinner, my best friends from Nicholls State and Ole Miss, and all the spiritual people I have encountered in all venues. The very best of these know who they are.

There is no finality to my acknowledgements lest I miss someone. The representatives, editing staff, and publishing staff at Westbow Press were instrumental in convincing me that putting my thoughts out there may be a worthwhile endeavor, and may help people understand and build on how the world works.

Introduction

Science begs—no, demands—inquisition, and it must be challenged. Theories are no more than stories we tell from our understanding of the natural order. From a tiny, short-lived, human perspective, we surmise how something exists, build on that assumption, and base our most fundamental conceptions on that information.

To understand something such as climate variability, we must understand the context of whence we came. Our tilted, rotating, orbiting Earth; revolving solar system; and spinning galaxy owe their eccentricities to events deep in the cosmos that have evolved over space and time.

To understand our planet requires a thorough understanding of the structures, forces, and interactions that shape the universe. Unfortunately, we can't even be sure most of the universe still exists. We are forced to understand our Earth, its workings, and the rest of the universe from the inside of our planet out. Our understanding of the forces that shape our planet is minuscule in size and time compared to the scale of creation.

The biggest truth that stands above all logic and reasoning is that this Earth is the only planet in all creation that contains life. We may surmise that life exists elsewhere, but it is unknown to us. We may not be the center of the astronomical (theoretical) universe, but we are the center of the known universe because it too exists only to us. From our perspective, we can see only so far in all directions, and the distance we can see is the same in all directions.

We are also the center of the anthropological universe, the universe that contains life. All creation is molded around this miracle planet. Our positioning relative to our Sun has been determined by our positioning relative to the universe, and the perturbations in the universe have allowed Earth to evolve into a life-sustaining planet. This makes the universe our own.

Miracle Planet

As far in space as we can see and as long in time as we know, we have determined that Earth is the only planet in the wide expanse of the universe capable of supporting life. Thus, the universe exists only to us. We invented time and space. Our knowledge of the universe controls everything we know about our existence and relationship with our surroundings.

We observe, speculate, and measure to determine our relations to the things around us that are controlled by *the* higher power that created them. In one sense, it might be better if Earth were not the only sphere of life; then, our circumstances would be easier to explain. Either the universe came into being somewhere in time, or it already existed.

We are but a tiny, weak speck of life. The human form is about one one-hundred-thousandth of a percent of the diameter of Earth. It's easy for us to measure distance and time in the realm of our Earth, but measurements grow more complex as we look at the stars. It takes about eight minutes for the light from the sun to reach Earth, which means that the Sun we see is not as it exists now but as it existed about eight minutes ago.[1] Compared to the known diameter of the universe, Earth's diameter is about 1.4 quintillionth of a percent.

Here's where we stand by the numbers (in feet): humans—five; Earth (diameter)— 42 million;[2] universe (known diameter)—3,000,000,000,000,000,000,000,000,000 (octillion).[3] We are like ants—or even smaller creatures—crawling on the bottom of the abyss and trying to visualize what lies above.

Our tiny existence relative to the scale of creation makes us a most random occurrence. If God had created the universe, it is only humankind that perceives it. Everything we know about the universe comes from our early inspirations, scientific thought, theories, observations, and measurements. As humans, we are in sole possession of this knowledge. The concepts of space, time, scale, measurement, and change were all invented and developed from our perspective.

There is no miracle greater than the miracle of our existence. When we define a miracle, as, from WordWeb®, "any amazing or wonderful occurrence; or marvelous event manifesting a supernatural act of a divine agent," we couch the definition in human terms. From our perspective, we consider only those things that defy the natural order to be miracles. The universe may be the first miracle of creation, but Earth is more of a miracle than that. It circles the Sun at just the right distance and with just the right tilt and rotation to allow it to support life. Due to its relationship to the other planets and stars around us, it has just the right gravity to support an atmosphere and climate that enable life to thrive.

Only on Earth, which is an infinitesimally small part of the universe, does the natural order include life, the biggest miracle of all. Humans and other forms of life on our planet are but temporal specks of matter compared to the space and time occupied by the universe. All nature, as we define it, is but a temporal cloud of existence compared to the existence of the universe. Humans are part of nature. Without humans, nature would have no need to exist except for existing on its own. It would be unknown.

Drukell B. Trahan

We are very small compared to the universe. The known universe has a diameter of approximately 93 billion light-years (each light-year equals approximately 6 trillion miles) and is 13.7 billion years old.[4] This is relative to our earthly (human) perspective. By comparison, Earth's diameter is only about 8,000 miles, a fraction of 1 light-year, and it is estimated to be 4.5 billion years old.

Our human experience is limited to what is within our reach. The nearest star system to our own is about four light-years, about 24 trillion miles, away.[5] At the speed of sound, a speed we actually have been able to achieve, it would take us about 4 million years to get there.

Universe

The universe extends from us to a distance that we can see based on our understanding of space, time, and the speed of light. The known universe extends about 45 billion light-years from Earth; the farthest light we can see with our telescopes was flashing 45 billion years ago and is 1.5 octillion feet away.[3]

The nearest star outside our solar system is four light-years away; at the speed of light, it would take us four years to get there. It also means that we see the star as it existed four years ago. We know it exists because we have seen it well within the time frame of human history. Other stars we can see have existed within the time frame of human history, or in our case civilization, which we could estimate conservatively at 7,000 years.

Beyond this observable limit is the edge of our galaxy, 25,000 light-years away,[6] and the nearest galaxy to ours is about another 25,000 light-years away.[7] Our own Milky Way galaxy is approximately 100,000 light-years in diameter.[8] We've had the ability to measure these distances for only about 200 years. Our measurements of the stars are based on complex calculations and assumptions that shape our understanding of an expansive, incomprehensible system that defines everything we know about our existence.

We know only that those objects in the universe in a conservative 7,000 light-years of Earth have existed since we started observing them. We are the center of our universe out to about 7,000 light-years; that much we know for certain. Objects farther than 7,000 light-years away are only assumed to continue to exist; they may be much different now than as they appear to us. However, we believe that our solar system and its motions are influenced and preserved by the motion of the galaxy, which in turn is influenced and preserved by other objects in the universe. On top of that, the universe is expanding and churning at a very slow rate that is decreasing with time.

The spherical nature of the objects in the universe and their motions appear to be part of the natural order that extends through all creation, including the smallest atoms.

Everything we know is made up of atoms. Always depicted as a rotating sphere with smaller orbiting spheres, an atom is nonetheless a spherical system. Molecules and cells are made of atoms that combine to form minerals and life forms. All this makes up Earth, a rotating planet surrounded by an orbiting moon.

The next spherical system consists of a rotating sphere (the Sun) surrounded by orbiting spheres (the planets). Galaxies (also spherical) are made up of many solar systems. All this makes up the universe. If the expanse of time and space continues, the universe would also be a spherical system and a part of the next larger system.

The rotations of spherical systems appear to have eccentricities of motion that emanate from a common source. Atoms degrade. Earth wobbles. The planets spin around a pulsating Sun. This rotation and eccentricity extend into each object. The mantle of Earth rotates, and the atmosphere roils. Again, the speed of rotation is decreasing with time though at a very slow rate.

The similarities of the spherical systems, their rotations, and their eccentricities are easy to understand using the concepts of cause and effect. For every action, there is an equal and opposite reaction. If you heat a pipe at one end, the heat is transferred through the pipe to the other end although gradually and in decreasing strength. Snap a rope, and the energy is transferred through the rope to the other end in wave form and in decreasing strength. When you observe an effect, you can determine its cause, which will be where the energy was applied. The orbiting nature of the objects in the universe—down to the smallest atoms and perturbations in energy—provide evidence that the energy source at the beginning also was orbiting and pulsating. This observation is not as important as is understanding these phenomena as parts of the natural order; everything in our sphere of existence is controlled by these perturbations of energy, including the action of the Sun, the wobble of Earth, and the roiling atmosphere.

Earth orbits the Sun at about 67,000 miles per hour and rotates at about 1,200 miles per hour.[9] At the same time, Earth is moving from within at a very slow rate. It consists of a solid inner core, followed by a liquid core, viscous mantle, and crust. Layers of decreasing density, pressure, and temperature from the inside out are moving in ways relative to variations in density, pressure, and temperature.

It doesn't stop there. The oceans and other waters make up another layer—the hydrosphere—with particular properties of density, pressure, and temperature—followed by the layers of the atmosphere composed of gases, including water vapor, which have properties again based on density, pressure, and temperature.

Our Earth is like any other body of rock in the solar system except for the liquid and gaseous interactions that sustain its life. Earth also contains one other all-important exception: layers of carbon. Its atmosphere, oceans, and land interact with each other to support a circle of life that as far as we know exists nowhere else in the universe. As far as we know, by sampling, carbon is found elsewhere (moon rocks and meteorites) in only trace amounts. Carbon is measured spectroscopically only on other planets and the Sun.[10]

But of all the special properties of our planet, the atmosphere (gases) caused by the interactions between Earth's inhabitants (the biosphere) and its configuration of solids (the lithosphere) and liquids (the hydrosphere) is most necessary for our survival. Each sphere has properties controlled by the energy stored up since creation and is found in the spinning and heat of the objects that make up the universe down to the smallest atom.

However, the atmosphere protects us from heat and the occasional celestial object, sustains life with its gaseous interactions, and impresses us and confounds us with its energy. Of course, our particular atmosphere would not exist if it were not for the interactions between it and the other spheres.

The atmosphere is like any other collection of surfaces that roil around the planet. Though we breathe and thrive under this atmosphere and seem immune to it at times, it is nonetheless a physical object that exerts its own pressure. If the universe is expanding and the rotation and energy of its objects is decreasing with time, so are the other rotations within. This would include the energy in the systems controlling the atmosphere and climate, although this is happening at a very slow rate.

Randomness

Think of humanity and all other life on Earth and Earth's ability to sustain us as *the* most random occurrences. Along and among this span of time and space we call the universe, this life is all that we know exists; it consumes the gases and elements that make up the atmosphere and environment that sustain it, transforms the gases and elements, and returns them to their surroundings. The atmosphere and environment in turn consume life.

We know of many rocks like Earth that cannot sustain our carbon-based life based on our current knowledge. This is a mystery humanity has been trying to understand since it came into existence a short 7,000 years ago; that is out of the 14 billion years that the known universe has existed—only 0.00005 percent of time as we know it.

The solar system in which Earth is has a diameter of approximately 5.5 trillion miles, about 0.000000001 percent of the universe.[11] Considering such an immense span of time and space, how random is the existence of life?

Because we are so random, we are either infinitesimally important or even more infinitesimally insignificant. Let's just say for argument's sake that we are important. That would mean that our randomness is no accident and that our relationship to the universe as we know it is as both consumer and steward. It is our responsibility to care for the universe and harness its energy.

Our relationship to the universe starts with our relationship with the nature that surrounds us and has been entrusted to us. We share Earth with the rest of carbon-based life that has to be infinitesimally important as well. All carbon-based life consumes, processes, and emits the gases and other elements that surround us; this is a natural response to our surroundings. Even still, we are the dominant species because of our ability to reason and invent technologies that go beyond the abilities of the rest of nature. Bugs may be able to build what they need to survive, but we go beyond that and increase our levels of comfort, entertainment, and longevity. This is our right and our responsibility as the dominant species.

Life, as insignificant as an accident, does not have the same responsibility or right to harness the power of the universe or to care for it. We would exist for the benefit of our surroundings and therefore have no purpose other than to survive as is the case with the rest of nature. But owing to our nature and our abilities to reason and invent, there is more to our existence than survival alone. Until we find life elsewhere, this has to be the basis for our understanding. As the soul survivors of the universe, we are entrusted with the harnessing of its energy for our evolution.

Change

All the change around us is evidence of the nature of our relationship to the universe. Ever notice that change appears to be mainly in a decreasing direction? Everything we see is subject to change; it goes from order to disorder. Metals corrode. Water evaporates. The biomass ages. These processes take place due to the application (or loss) of energy. This is energy in the form of electrolysis in corrosion, in the form of heat for atmospheric processes, and in the form of biochemical processes when it comes to aging. The only way to turn either in the opposite direction is to apply more energy or take it away. Though we cannot perceive it on its scale, the amount of energy available for these processes must be dissipating over time. Therefore, energy may not be constant as once believed.

Relative to the size and age of the universe, our units of measurement are infinitesimally small. We know the universe is expanding; this is one of the few things we can measure precisely. It is the measurement of the change in the position of objects relative to each other. We cannot see the edge of the universe because we cannot see that far back in time.

If the universe started as a big bang, its energy has been dissipating over time as it expands. This is practically imperceptible to us owing to the amount of energy stored in the universe, and we cannot control this change. Understanding that the energy stored around us from that initial burst is going in a negative direction, however, is something we can use to explain the change and adjust to it either intentionally or because of some other activity. The only control we have over change is our reaction to it and how it affects us.

If the universe is finite, it has a beginning and an end. This makes everything else we think is infinitesimal also finite but still exceedingly and incomprehensibly large. Just as the universe has a beginning and an end, so does time, energy, and matter.

Keep in mind, however, that the big bang is a theory. Another possibility is that the universe is infinitesimal.

Like Ants on the Bottom of the Abyss

Remembering that this measurement is relative to us, that makes the edge of the universe incomprehensible. We measure time. We invented time. Relative to us, time is also infinitesimal. Be aware also that we invented ways of measuring time and space. By design, our numerical system is infinitesimal, so that makes everything we measure potentially infinitesimal. There is always a larger positive number and always a smaller negative number.

What about the opposite direction—toward the infinitesimally small? We split the atom into parts; can we split those parts into smaller parts and so on? As for time, there is no opposite, negative direction.

Everything is in a state of change in a seemingly decreasing direction. The universe is expanding, time is increasing, and energy is apparently being used up. However, relative to us, this system is infinitesimally large in all its capacities. The fact that it is changing this way is relevant but only to us. This is not necessarily a doomsday perspective. It is not even necessary that we comprehend it completely due to its scale. This change is not necessarily to our detriment as long as we remember that it is relative to us. It is infinitesimal to us, and we are but an infinitesimally small part of it.

It is also our responsibility to manage the forms of energy that make up our environment in this small part of

Drukell B. Trahan

the universe independently and in unison with change. Just as it was our manifest destiny as a civilization to discover and consume the land that extended west from the cradle of civilization, it is also our responsibility to use energy in responsible ways for our technological advancement. As far as we know, we are the only sphere of life and intelligence in the whole expanse of time and space, and we have the responsibility and destiny to do so.

Measurement

What average temperature is Earth supposed to be at right now? We know from our recordings of temperature for approximately 150 years that Earth is in a warming trend. Earth has experienced warming trends before; to what extent is the current warming trend different? To determine that, we rely on the historical records of ice, rock, and trees.

Because of the variability in temperature estimated over the ages, we do not attempt to compare the global average temperature now with previous global average temperatures. Rather, we compare the change in global average temperature over time.[12] To measure the degree of warming historically, scientists rely on a measure of the change in temperature relative to a global average temperature that is statistically calculated using monthly observations collected since 1880, but that is a miniscule amount of time compared to the age of the universe.

We need a baseline to measure change. We assume there is a temperature at which to assume equilibrium or zero change. We can't measure anything without a reference point whether it's zero or zero change.

Take for example elevations of the land or sea. We know the elevation of the land relative only to the sea surface and vice versa. If one changes, so does the other. Elevation and distance are relative measurements. To measure either, we have to assume that our reference points are not moving. Fractions of our measurement units may not matter in a moment, but they will if we come back after a time to measure any change. Larger units of measurement give us a greater level of comfort with more consistent results. Also, distance is subject to change; everything is moving.

In terms of temperature, absolute zero is the point at which there is no amount of heat; it is the coldest theoretical temperature possible. We also measure time from a theoretical zero that comes after the absence of existence. Unless we determine how to decrease time, time will always be moving forward. It has a beginning and an end if not infinitesimal. This is how we measure time. Temperature, however, can be either positive or negative relative to zero; that is by humanity's design.

Temperature, elevation, and distance are relative measurements. We developed relative measurements based on observed relationships. A foot became a foot,[13] 360 became the days in a year and degrees in a circle,[14] and twelve became the months in a year, hours in a day, and inches in a foot.[15] Zero degrees Celsius (C) is the freezing point of water, while 72 degrees Fahrenheit (F; 22 degrees C) is a comfortable temperature. Maybe the most comfortable temperature should be zero. Extreme temperatures on either end are extremely uncomfortable or deadly; it can be either hot or cold as hell. This is not the case for concentrations as in CO_2 and dust, which are measured relative to an absolute zero where the substance is absent.

In quantifying changes in temperature over time and comparing them with other changes, we are comparing relative measurements (temperature, sea level, time) with absolute measurements (CO_2, dust). We know where zero is for the absolute measurements; we do not know where zero is for the relative ones except maybe theoretically. Does the absolute zero temperature actually extend to the beginning of time when heat was absent?

Drukell B. Trahan

This would mean that before the beginning of time, when heat was absent, the temperature started out at about -273 degrees C (-459 degrees F). Again, this whole system of measuring temperature and time is by our design; we don't know the beginning or the end.

Today, the average global temperature is about 14.5 degrees C (58 degrees F),[16] and it can vary between -89 degrees C (-129 degrees F)[17] and 57 degrees C (134 degrees F).[18] According to the geologic record, it would have been about 1 degree C (34 degrees F) warmer about 5.5 million years ago and would have varied between -95 and 168 degrees F.

Scale

When you see something that was built to scale, is it really true to scale? For example, the door handle on a scale model of a Lamborghini may be in direct proportion to the door per the proportion when measured on a real Lamborghini and even be manufactured with the same metal and paint. However, the molecules of metal and paint in the model's door handle are not in direct proportion to those in the life-sized version. This gives the small-scale version different properties than the real Lamborghini has.

Another example more relevant to the study of natural systems would be the construction of a river model at a smaller scale than the actual river. The dimensions of a sand bar can be scaled down to match the ratio of the river width and length. The sand grains too might be scaled down to very fine particles but the water molecules cannot be scaled down. This means that the forces acting on the water and sand at the smaller scale cannot be proportional to the forces acting on the river at actual size. They are much larger in comparison.

Ever notice that a one-foot section of plumbing pipe cannot practically be bent but a fifty-foot section can hardly be prevented from bending? Longer sections of pipe have different properties of strength than do shorter sections due to different and larger forces (gravity, or the amount of flexibility, for example) acting on it at the larger length.

What does this have to do with the concept of climate variability for example? Climate variability is studied on a global scale. The rise of the sea level, one of the supposed effects of global warming, is confounded by the variability of Earth's surface due to subsidence, postglacial uplift, and earthquakes. It is difficult to imagine that the land surface usually thought of as being stable and immobile actually floats around on a sea of denser rock. Similar to the pipe example, given the relative size of Earth, rocks have different properties of strength as their size increases. Add to this that Earth may be increasing in size with time, however at a very slow rate.

Time is also a factor. Over hundreds of thousands to

millions of years, the Earth's crust bends and deforms, is subducted and lifted at spatial and temporal scales that are hard to conceive. The larger in scale an object such as Earth becomes, the greater the forces acting upon it; when it comes to the universe, the forces acting on it are even greater. This is also difficult to conceive on our relative scale. Remember that time is a term we gave to all existence and our way of measuring it however long that may be.

The pipe mentioned earlier took on a marked change in its properties of strength over a small change in scale. The diameter of Earth is 8.4 million times the length of a five-foot person. How much more of a change would we expect from our perspective over this degree of scale difference? Note that the scale of time also increases along with the change in scale of size.

The diameter of Earth is 132 times the thickness of its atmosphere—not a very large scale difference. That is why the atmosphere at the scale of Earth is influenced by the same large forces and has the variability it does over time. If Earth bends and deforms spatially and temporally, how much more would the atmosphere do that given its relative density?

People

As we understand them now, people became people with the start of the first civilization depending on whom you ask between 5,000 to 7,000 years ago. No one is in disagreement on this general time frame whether you believe in creation or evolution. This beginning was dependent on our ancestors becoming intelligent enough to develop societies, infrastructure, bureaucracy, monetary systems, religions, and the written word to name a few attributes of civilization. We know little about humans before that time except for what we can gather by the study of the fossil record and genetics. The appearance of the first humans prior to recorded history is theorized based on fossil evidence, genetics, anatomy, and intelligence (for example, the use of tools).

Drukell B. Trahan

If you believe humans evolved from apes, the theory of evolution, the first humans appeared approximately 160,000 years ago. One has to wonder then why it took another 150,000 years for a civilization to develop. A more likely scenario is that we ascended from apes but didn't become fully human until we developed reasoning and morality. Something happened to spark a change that separated humans from the rest of the animal kingdom and establish us as the dominant species. Evolutionists explain it with the great leap forward theory and believe this happened about 50,000 years ago. But again, why did it take even another 40,000 years for a civilization to develop?

Since the level of oxygen in the atmosphere has built up over time, it seems that during the early days of human evolution, it may have been lower. Perhaps the increase in human mental capacity was related to an increasing percentage of oxygen in the atmosphere. The longevity of certain biblical figures may be related to early levels of oxygen being lower; with an increase in oxygen came higher intelligence but a shorter life span. Also, that it didn't affect other animals the same is one proof we were "destined" to become the dominant species.

To me, this is where God comes in. Of course, he was

always there, but about 7,000 to 10,000 years ago, he gave us reasoning and morality, and from there, we went on to develop civilizations and a written history, and that's when we became human. God gave us reasoning and morality, but it seems he didn't distribute it evenly among us. Some of us have greater reasoning abilities and a stronger sense of morality than do others. That's because he also gave us free will and independence.

He didn't give reasoning and morality to the other animals, but he did give them free will. So free will isn't special; it's a bane. On the other hand, our ability to reason and know right from wrong is special. As random as we are as a species, we are left to wonder why. What is our purpose on the grand scale of all existence? We are but an exceedingly small and incomprehensible part of it.

Manifest destiny was the belief that a civilization had a right and responsibility to discover and consume the land that extended westward from the cradle of civilization. The civilizations in the cradle were ultimately no more than humanity's own design because of its inability to accept that it was one race, the human race. Its desire to spread civilization around the world displaced and eventually absorbed the tribes that once spread across the lands.

Drukell B. Trahan

    Are these absorbed cultures any worse off now than they would have been? Of course, some weren't given a choice between slavery and freedom, and that may have been a bad influence, but the tribal existence present among these cultures and some continuing to this day doesn't take advantage of the many conveniences and comforts that science and technology offer. What would the world have come to if tribes had been allowed to become the dominant societies? We should look at a global civilization as the soul of the universe. As far as we know, we are the only sphere of existence in the immense span of time and space we call the universe.

God

God exists. What our universe encompasses is just too incredible to believe otherwise. One of the main arguments against God is that there must be other beings out there somewhere given the universe's incredible size. Otherwise, it's because of the size and age of the universe that there probably aren't. By now, after 4.5 billion years of Earth's history, we should know.

It's amazing that our Earth cooled to exactly the right temperature to support life and adjusted itself to allow us to exist as the dominant species. Our world is in just the right orbit relative to the Sun and other planets to be capable of sustaining life. We breathe oxygen and give off carbon dioxide while trees do just the opposite. Perfect, except for the fact that the oxygen we breathe is actually what eventually kills us.

It is also easy to accept and understand how ludicrous is may seem to believe in a God, Son, or Spirit who have been unseen for so many years. After all, it's been about 5,000 years since anyone has seen God, and it's been about 2,000 years since anyone has seen his Son or the Spirit. But that's 2,000 years out of 4.5 billion, a drop in the proverbial bucket of time. We don't have enough data.

Isn't it also as or more ludicrous to believe that we ascended from a primordial ooze through other primates? This is another theory that has not been proven. Does any evidence exist in the geologic or anthropologic record other than our resemblance or relationship to other life forms? Why wouldn't God have used the same basic building blocks to create everything? There may be some truth to evolution and natural selection, but we don't have enough data to prove our ascension from nothingness by chance.

Although we are part of nature, we do not follow the laws of nature. That sets us apart from the rest of the animal kingdom as God intended. Our laws are based on God's laws. It's a good thing we were given God's laws and don't have to follow nature's laws or we'd be relegated to living like the other animals. Those of us who follow God's laws could be earning our places in a new world where no laws, not even the laws of physics, would be necessary.

So what there is here on this Earth is all there is—God's "special" creation with all the rules he has provided, some written, some for us to discover. God gave us Earth to take dominion over. He didn't want us to not discover electricity or not to take oil out of the ground for our advantage. He does, however, want us to care for everything he has given us and use it sensibly.

Bandwagon

We can think for ourselves, but is something true because a bunch of people agree it is, or is something not true unless a bunch of people agree?

Consider man-made climate change (MMCC) for example. A small group of scientists came up with the concept and got a bunch of other scientists, politicians, and Hollywood actors to agree with it. How many of those have actually looked at the data? The atmosphere and thus the climate is probably the most significant attribute of an Earth that sustains life; it must be understood and managed according to the natural order.

The problem with peer-reviewed scientific research is that out-of-the-box ideas aren't accepted if they don't fit into the very small picture of what is accepted. However, when early scientists were coming up with their ideas, there were few if any peers, and still, their ideas turned out to be the bases for everything we believe today.

Supposedly, numerous scientists accept MMCC. They and everyone else should understand that people, including scientists and politicians, are human and as such tend to follow each other. There may be a smaller group that does not accept the theory, but then again, originally, the group that did not believe in a flat Earth was small too.

Climate science is multidisciplinary. Many factors influence the climate—Earth's position relative to the Sun, the composition of the atmosphere, and ocean currents to name a few. Changes in climate whether small-scale, large-scale, short-term, or long-term create what we refer to as the weather and change it spatially and temporally.

The list of scientists who believe in MMCC includes physicists, climatologists, meteorologists, chemists, biologists, and oceanographers but very few geologists.[19] The list of scientists who deny MMCC or doubt its magnitude include a few physicists, climatologists, meteorologists, oceanographers, but many more geologists.[20]

This is telling. Physicists base their studies on either the very smallest (atom) or the very largest (universe). Much of what they do is relegated to scientific theory, and much

Like Ants on the Bottom of the Abyss

of what they work on is either invisible or untouchable. Climate scientists study the interactions and changes that occur at the molecular level, say between the Sun's radiation and atmospheric gases, or they study the forces involved in the winds and waves and in the motions of the planets.[21]

Chemists, climatologists, biologists, and oceanographers each study one aspect of Earth. Chemists look at the elements and compounds involved in climate interactions. Climatologists study the atmosphere, and oceanographers study the hydrosphere (oceans), both looking at patterns and formulating models that help them understand ongoing and predict future events. Biologists study plant and animal life and look at their interactions with the environment; the most important one here, obviously, is the effect humans have on it.

Geologists study Earth or at least more of it than any one of the other disciplines. The compositions of rocks, sediments, and water (including ice) are dependent on their environment of deposition, or how and why they were laid down in the patterns they form, including climate (weather). The various layers and their patterns of deposition reveal the changes in climate that have occurred over time. It doesn't matter what caused the changes, only that changes have occurred and can only be natural at least up to a point in time. To understand the history of Earth and its climate stored in the rocks, sediments, and water, geologists strive to understand recent environments, which include the sediments being deposited now.

Unfortunately, the interactions studied by each discipline would occur whether climate change was natural or man-made. For example, the claim is made that increasing CO_2 causes warming, but warming oceans release more CO_2, so is the increase in CO_2 the cause or the effect? There are many examples of this paradox. Humans interact with the environment in a natural way. Who is to say at which point our natural interaction ends and our human interaction begins? We are supposed to be industrious.

The main line of evidence for MMCC that is the basis for all the climate science research in physics, chemistry, biology, climatology, and oceanography is not that Earth is warmer than it's ever been but that carbon dioxide levels are higher. The history of climate variability over the past 450,000 years is contained in one graph developed through the analyses of Antarctic ice cores (see page 62). The graph shows the periods of cooling and warming along with decreases and increases in CO_2 and dust levels corresponding to the waxing and waning of the ice ages. Actual measurements of CO_2 levels since about 1950 have been tacked on to "prove" that CO_2 levels are increasing at faster rates than they did during previous warming periods.[22]

The MMCC scientists have made some assumptions to come to their conclusions. The first is that Antarctic ice is always laid down the same way and that there are no changes in deposition or compaction rates. Another assumption is that the CO_2 gas measured in the ice is all from the layer that is being evaluated and that the CO_2 bubbles do not migrate between layers. These assumptions introduce potential error.[23] There are margins of error in the data from recently collected data (since 1950) as well, which can vary due to point sources of CO_2 and atmospheric circulation anomalies.

Also, certain nuances of data should be considered when viewing the graphs. The data is graphed on a vertically exaggerated scale that makes the variability look large when actually the supposed change is about 100 parts per million, or 0.01 percent. That's like having a teaspoon of sugar that's missing eighteen grains. Adding the missing grains would result in an imperceptible change.

Geologists have been studying climate variability longer than they have the historical record climate scientists use to insinuate MMCC. It is the very basic tenet of geology that the history of climate variability is stored in rocks,

Drukell B. Trahan

sediments, and water, including ice. Data collection and interpretation is not an exact science. Assumptions always call into question comparisons with present, real-world observations. Climate scientists and politicians know three things: that the climate has changed, is changing, and will change. No one, not even geologists, know what's going to happen next, or to what extent it will happen, or the interactions that may influence climate variability. Even meteorologists can't always get their predictions right from day to day.

Man-Made

It seems that everywhere you turn, a new study verifies that Earth is warming, that the ozone hole is getting larger, that carbon dioxide levels are increasing, that polar ice is melting, that sea levels are rising, and that ocean currents are being altered. And it's all because of human degradation and destruction. And there are a few opposite studies that may indicate more positive results, but these are given short-shrift; such as studies that indicate at times the ozone hole has gotten smaller.

"It's the worst it's been in recorded history" has become the mantra for the media to apply the label *man-made*. However, in all the sensationalistic news, there is never an explanation that recorded history is infinitesimally small compared to the histories of Earth and climate variability.

So why do MMCC enthusiasts insist on crediting human capitalism with our current warming climate? Could it be that everything we do burns up oxygen and turns it into carbon dioxide? That's not unusual; every species of animal does that. Humanity's contribution of CO_2 to the atmosphere by industriousness is estimated to be 3 percent by volume. The natural contribution to atmospheric CO_2 is approximately 24 percent by volume.[24] If we weren't so industrious, we'd probably contribute only a small percentage (say about 3 percent) to this. So as a species, we contribute about 6 percent to the total. How much of the 6 percent is due to agriculture and other wastes we generate that would be present no matter how industrious we were unless we stopped producing and harvesting in addition to discontinuing certain practices such as the manufacture of plastics?

The percentage of CO_2 in the atmosphere is about 0.04. So when a climate scientist argues that it is increasing, it is from a very small amount to a very small amount. CO_2 levels in the atmosphere have increased previously as measured in sediments and ice cores to levels that we are seeing now. Remembering that historical levels are measured with a varying degree of accuracy, they could have been somewhat higher or lower years ago. Supposedly, however, CO_2 levels are increasing at a faster rate than they would naturally due to our extra contributions. The natural rate of increase again, however, is subject to measurement error.

So just how accurate is the science of MMCC? According to the Intergovernmental Panel on Climate Change (IPCC), the warming Earth, spreading ozone hole, increasing carbon dioxide levels, melting polar ice, rising sea levels, and shifting ocean currents coincide with changes to the levels of gases being contributed to the atmosphere by humanity.[25] Funny thing is we've only had the ability to measure temperature, sea level, ice and currents about as long as we've been able to measure the concentrations of the gases. This doesn't mean that the gases have caused the warming and other changes, only that we've been measuring them at the same time.

Correlation does not imply causality. Earth has been warming naturally since the last glacial maximum and more significantly since the Little Ice Age.[26] Though the warming and changes in ice and sea level may be related, the relationship between human-made gases and warming is only coincidental at best. Carbon dioxide, methane, nitrous oxide, and miscellaneous other gases are referred to as greenhouse gases. The miscellaneous gases include chlorofluorocarbons (CFCs) and the like, which are the only gases not produced naturally, but they account for only a fraction of a percentage of the atmospheric gases. The concentrations of carbon dioxide, methane, and nitrous oxide produced by human industry account for approximately 4 percent of the total atmospheric concentrations of these gases. Natural sources account for approximately 23 percent, and the remaining 73 percent were present in the atmosphere prior to industrial development and had to have built up naturally.

Earth is warming and will cool again naturally as it has done many times. It may coincide with changes in atmospheric gases as it also has previously. Whether there is a cause and effect between climate variability and the gases, the natural contribution far exceeds that extra part contributed by humanity's industriousness. And insomuch as we are a part of nature, what part of our industriousness is not also natural?

Forecasting

How much of our perception of climate variability is due to an increase in our understanding of the subject? How much of the apparent increase in severe weather is due to our increasing ability to predict with increasing accuracy and timing the arrival of severe weather events?

At one time, tornadoes and other severe weather events arrived unbeknown to the people in their paths. As recently as only approximately 150 years ago, tornadoes and hurricanes would wreak havoc and be recognized for their force or pass through unpopulated areas unnoticed and miss being catalogued at all. Then came our abilities to measure temperature and pressure, to monitor larger expanses of Earth, and to transmit information over longer distances all with increasing speed.

Informal weather forecasting began approximately 3,000 years ago with the Babylonians; it proliferated through many cultures and related changes in weather to astral signs, lunar phases, and wind patterns.[27] Although some of the relationships have held up through the ages, many have not.

Repeated changes in weather and the timing of specific events undoubtedly held sway with patterns that had as their basis the arrangement of the seasons. Inhabitants of the world knew that a particular event had a greater likelihood of occurring at a particular time each year, the year being defined by astral and lunar relationships and patterns. Extreme events were hardly even recognized. It's one thing to be able to measure wind speed but another to feel it. The death and destruction from a storm was bad regardless of the storm's intensity.

With the invention of the telegraph came the ability to warn others of bad weather, and the first boost in the ability to predict coming storms allowed people to prepare for them and measure their intensity. Then came greater advances in technology and a greater ability to sense and visualize the patterns that made for the changes in weather and climate.

In a short 150 years, we have gone from spreading

the news of coming storms by horseback to seeing instantaneously by satellite the movement of storms across continents; with that came the ability to measure storm intensity using barometers and anemometers rather than human perception alone. In a short 150 years, we have gone from having very little notice of a storm approaching to having a few days' notice and from having little knowledge of its intensity to having much more knowledge of that.

This is not climate variability caused by humanity but a change in our awareness of weather and climate variability, frequency, and intensity. The locations and timing of patterns have become more visible at an increasing scale. We aren't having more-severe weather but more and earlier warnings when severe weather is approaching and more information regarding its intensity.

Natural disasters have always affected life on Earth. Humanity is the only life form that is aware of the inconveniences caused by natural disasters. Though the odds against any human being affected by a natural disaster in a lifetime are large, these odds should have decreased due to technology being able to predict the onset of a flood, tornado, earthquake, fire, or hurricane. However, as our population grows, a larger percentage of the habitable land is occupied by humans and affected by technology. We are

Drukell B. Trahan

more likely to live, work, and play in some area of Earth prone to natural disaster, and natural disasters can occur anywhere. They are an effect and a cause of the atmosphere in an interaction between the perturbations in energy from the original source.

History

Every bit of the climate science used to associate climate variability with human activity is measured using data collected during the past 150 years at best. For comparisons with the historical record, if available, scientists rely on extrapolations using indirect methods. The ability to associate climate variability with changes in geographical phenomena is dependent on our ability to measure these phenomena on a global scale. Obviously, data collected early on is considered to be sparse and relatively inaccurate compared to what we can collect now. Improvements in technology, e.g., satellites have led to increases in the amount, distribution, and accuracy of data with time.

Temperature has been measured globally the longest—for about 150 years—and data from developed countries is the most complete. Other parts of the world haven't contributed much to the global temperature record until recently. Prior to the advent of accurate temperature measurements, scientists relied on data from ice cores and otherwise from the geologic record.

Also measured starting about 150 years ago were sea levels and ocean currents. The most accurate and global measurements of ocean currents didn't become available until the use of satellites in the midtwentieth century. Similarly, measurements of ozone on the ground were collected for approximately eighty years until the use of satellites enabled the measurement of atmospheric ozone (i.e., the ozone hole) about fifty years ago.[28]

Atmospheric carbon dioxide has been measured for approximately sixty years. Data from ice cores and other geologic methods are used to extrapolate atmospheric carbon dioxide levels as far back as 500,000 years.[29]

The jet streams have been measured and mapped for only about seventy years now. Studies extrapolating the positions and intensities of the jet streams as far back as 8,000 years have used oxygen isotope data from caves and lake sediments.

The extent of polar ice has been measured since about

1600, but the most complete records aren't available until about 1750, or for about 250 years.

Given this information, to be able to correlate all these variables with climate variability, specifically **MMCC**, the best matching data are available only for the minimum data set or about fifty years. In addition, each of these phenomena can be associated with other causes, for example, Earth's temperature and the jet streams due to variations in tilt of Earth, sea levels due to changes in the elevation of the land surface, ocean currents due to changes in the configuration of the land masses, and concentrations of atmospheric gases due to all natural processes.

Historical

Wikipedia's entry on "Ice Age" shows temperature, carbon dioxide, and dust changes over the last 450,000 years. A few observations are notable.[30]

1. Previous interglacial (warm) periods (higher temperatures on the graph) have gotten warmer than the current interglacial (warm) period based on temperature. These cannot be man-made.
2. Some previous interglacial (warm) periods also had higher (or at least as high) carbon dioxide levels than the current. These too cannot be man-made.
3. Past measurements up to the most recent 150 years or so are based on estimates that may have some error, so the magnitude and duration of change may be slightly off.

4. The current interglacial (warm) period has lasted longer than previous interglacial periods; this may be partially due to the previous point. Changes in temperature over the last 150 years or so are based on actual measurements using thermometers, while previous changes are based on measurements using proxies such as ice cores and tree rings.
5. Looking at the last few thousand years close up, it appears that the average temperature is decreasing, which would be expected if we are tending toward a new glacial (cold) period (see below).
6. Higher dust levels correspond to the glacial (cold) periods and are low now.

So what has happened with temperatures during the last few thousand years? Wikipedia's entry for "Temperature Record of the Past 1000 Years" shows a general increase in temperatures during the last 1,000 years.[31] Even NOAA shows temperatures to be generally increasing during the last 2,000 years.[32] Ice cores may indicate that other analyses may not be what they seem—not all ice cores are created equal. One shows that temperatures have been trending downward (colder) for the last 10,000 years. It also shows temperatures increasing during the last 300 or so years, more noticeably during the last 150 years.

But remember, 150 years marks the extent to which we have "actual" temperature measurements. You can look at temperature records from just about anywhere and get the impression that temperatures are increasing for the last few hundred years or so, but you need to look at temperatures prior to that time, and just about every analysis agrees that temperatures have been warmer than they are now and may even be decreasing recently.

Climate Change 101

The climate is always changing. Humans may contribute to that, but that contribution is negligible and to some extent unavoidable. It's not that we should do nothing about it; we should, but to a limited degree. Our constant efforts to control things sometimes have negative consequences. If we were to swing into another ice age, we might be happy that we warmed Earth a little.

In all the sensationalistic news about MMCC, however, there is never an explanation that recorded history is infinitesimal compared to the history of Earth and the history of climate variability. In geological studies, we learned that Earth started as a molten mass approximately 4.5 billion years ago, cooled to a habitable climate, and has gone through periods of cooling and warming up to the present. It took approximately 500 million years for Earth to cool from about 18,000 degrees F to less than 100 degrees F on average considering Earth as a whole. Since then, it has gone through several periods of cooling and warming in the habitable range. At least one long cycle causes changes over millions of years. The cool periods included several ice ages. Warmer periods have tended to last longer. The most recent glacial period of the current ice age occurred approximately 25,000 years ago, and Earth has been in a warming period ever since. These cooling and warming periods are believed to be caused by changes in Earth's orbit that take place every 75,000 or so years. Other cycles caused by changes in the tilt and wobble of Earth take place over shorter time periods of 40,000 years and 20,000 years. There are also shorter cycles that we have been able to actually measure—an eleven-year sunspot cycle, a nineteen-year lunar cycle, the sixty-year Atlantic and Pacific oscillations, El Niño, and periodic but noncyclic events caused by volcanic activity.

We have been able to measure shorter cycles by taking actual reliable measurements for only about 150 years. We've had thermometers at weather stations across the globe and have been able to measure temperatures simultaneously or at least on a daily basis. The longer measurements show up in the geologic record, in sediment, rock, ice layers, and tree rings. Keep in mind that there is always some error applied to measurements we take using the geologic record or tree rings; dating a span of time using these methods usually gives you a measurement plus or minus hundreds or thousands of years.

We know that Earth has been warming since the last ice age, but we don't know when Earth may start cooling again or even if our current period of warming may be within one of the longer cycles. So given our 150 years of reliably accurate measurements and the age of Earth, 4.5 billion years, we know how the climate has changed for approximately 0.000003 percent of Earth's history. That's like looking at the weather for approximately 2/1000 of a second and trying to determine what the weather will be for the next twenty-four hours, and we know how much the weather can change in a day.

Even if we had 7,000 years of reliable data (the amount of time since the first civilization), that would be only about 0.0002 percent of Earth's history. If we take the biblical view that Earth is only 7,000 years old, we know only what the weather is like over 1/10th of a second as compared to a day.

One of the basic tenets of this science is that the history of Earth, including previous climates, is stored in the sediments, rocks, ice, and water. Remember again that this gives us a picture of how conditions may have changed over the 4.5 billion years of Earth's history, but the data are incomplete and estimated.

The Data

The data used to explain and interpret climate variability have certain statistical variabilities based on the accuracy of measurements and other causes of change to the parameters being measured. As mentioned, change is relative to a point in time when there was assumed to be no change.[33]

Take ice core data for example; they show swings in temperature, CO_2, and dust concentrations over 400,000 years.[34] What's most worthy of closer inspection is the amount of variability shown in the data over time and between parameters. Temperature data shows a great amount of variability over the most recent measurements compared to that measured during previous warm periods in history.[35] This is due to more-accurate measurements and easier calibration of the more recent temperature data.

Also, the recent (10,000-year) variability is on the order of approximately 2 to -2 degrees C (36 to 28 degrees F).

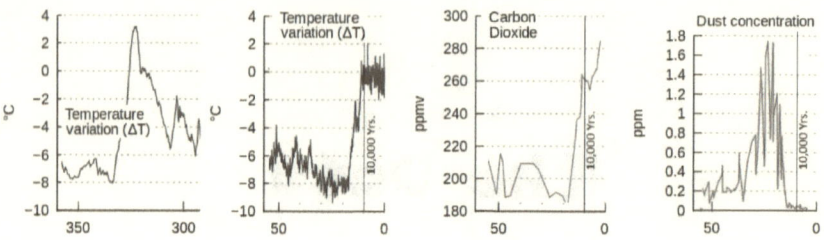

Temperature variability 300 to 350 thousand years ago and in the last 50,000 years along with the concentrations of CO_2 and dust

Note that the greatest variability in temperature and dust begins approximately 10,000 years ago coinciding with and (probably) preceding the dominance of humanity and particularly its use of combustion and an increase in its carbon footprint. CO_2 levels apparently increase with less variability during the same time.

From this, we see that while the more accurate measurements show CO_2 levels to be increasing, the temperature and dust data do not have a similarly linear relationship as CO_2. The geologic temperature record shows a variability of 0 to -8 degrees C (32 to 18 degrees F) over approximately 500,000 years.[36]

The 40,000-year and 100,000-year cooling and warming

Like Ants on the Bottom of the Abyss

cycles are associated with variations in Earth's orbit known as Milankovitch cycles. However, the overall historical trend is downward, or colder. Apparently, the current period of warming shown by the ice core data is at the warm end of a 100,000-year cycle.

Tree-ring data shows temperatures to vary between -1 and 1.5 degrees C (30 to 35 degrees F).[37] Note that while the frequency of warm events has decreased, the frequency of cold events has increased and that the magnitudes of recent warm events and cold events are about equal.[38]

A swing in the direction of temperature change from 35 to 30 degrees F appears to take place roughly every 400 to 600 years. According to the tree-ring data, we are due for a cold swing.[39]

The variability in temperature over the past 10,000 years should correspond to our dominance as a species over the same period. Rather, the variability does not show an increase but may even show a decrease in temperature over that time. Also, the variability appears to be dominated by the other forces that cause climate variability such as changes in the Earth's orbit.

There does not appear to be a relationship between increasing CO_2 levels and temperatures. Right now, temperatures are at the top of previously recorded temperature extremes based on ice core, geological, and tree-ring data. Based on the observed variability, there is as much a chance that temperatures will decrease in the future and by as much as 18 to 35 degrees F. Compare this to measured temperatures collected since 1960, which are the most accurate data we have and which are characterized by their own variability.[40]

To measure the change in temperature, the graph is zeroed on the average temperature from 1981 to 2010. This assumes that neither warming nor cooling occurred during that time, but the temperature varied between -0.5 to 0.5 degrees C (31 to 33 degrees F). This is within the range of minimum and maximum temperatures observed in the historical data from ice, rock, and trees.

Example

A rogue analysis of temperature data for the United States including Alaska and Hawaii is in the appendix. These data were selected because they are the most complete set of daily temperature records in the world as available from the National Weather Service. Graphs show the frequency distribution of daily record high and low temperatures in the twentieth century from weather stations across the United States.

The first obvious interpretation of the data is that many high and low temperature records were set during the twentieth century; this is just a result of the distribution of the data. Remember the earlier argument that we have only about 150 years of reliable measurements. Simply put, there is no data to show whether there were any record low or high temperatures prior to the twentieth century, so of course, all the record temperatures have been set during the last hundred years. We can consider the twentieth century only the baseline for record low and record high temperatures that rely on actual measurements.

The second obvious interpretation is that much variability exists across the United States including Alaska and Hawaii. There are more record highs in the Pacific Region (Anchorage, Honolulu, Los Angeles) and more record lows in the West South Central (Jefferson City, Baton Rouge), East South Central (Nashville, Mobile), and South Atlantic Regions (Raleigh, Jacksonville) excluding Miami. With this kind of variability across the United States, variability across other regions of the world would be expected as well. Even across the Arctic, it would seem logical that some regions of the Arctic might be experiencing higher temperatures and defrosting while others might be experiencing lower temperatures.

Jet Streams

The jet streams are caused by temperature variations with changes in altitude and latitude.[41] Air moves between areas of high temperature and low temperature from the land up and from the equator northward. As the air lifts northward and up, Earth, in rotation, drags the circulation with it to the east. Earth rotates at approximately 1,200 mph at the equator; that speed decreases with latitude. The jet streams range in speed from approximately forty to ninety mph and are normally the weakest near the equator.

We have only about seventy years of data of actual measurements of the jet streams starting circa 1948; this is about the same as what we have for measurements of ozone and less than our understanding of ocean circulation.

One study has used measured data during the last

thirty years or so to indicate that the jet streams rose in altitude and moved closer to the poles.[42] The jet stream in the northern hemisphere weakened while the southern hemisphere subtropical jet stream weakened and the polar jet stream strengthened. Despite every apparent intention to do so, the authors could not confidently attribute these changes to man-made activity.

Another study looked at data for the past 8,000 years using oxygen isotopes in caves and lake sediments.[43] The result of the study was that a decrease in the Sun's energy approximately 4,000 years ago resulted in cooler temperatures and an increase in the jet stream's "curviness." According to this study, the jet streams should be reverting to a straighter flow, but that the curviness is being exacerbated by climate variability.

Apparently, the two studies are contradictory. The short-term study using actual mapped and measured data surmised that flattening of the jet streams along with weakening and movement may be associated with climate variability, while the long-term, prehistoric data postulated wilder amplitude swings in the jet streams due to climate variability. Actually, the long-term study predicted without the influence of global warming what the short-term study showed.

Does this mean that the results of either study are wrong? Not necessarily. But it is also inappropriate to associate any changes in the position, curviness, or strength of the jet streams with climate variability; it may be appropriate to associate climate variability with changes in the jet streams. The cause of the jet streams is the planet's rotation on its axis and the degrees of solar radiation.[44] It stands to reason that changes in the rotation, tilt, and wobble of Earth along with solar cycles are the cause of variability.

Sea Level

As is the case with all the other phenomena climate scientists measure and attribute relative to the effects of climate variability (temperature, carbon dioxide, ocean currents, ozone), changes in sea level have not been accurately recorded for very long. It is also the phenomenon that causes the most alarm and is the most uncertain.

We don't ever actually measure the elevations of the oceans (the oceans are zero; see below), we measure the elevations of the land surface. The problem that had to be overcome is the lack of a baseline. For more detail on the science and history of elevation surveys, please see "Level Headed: A Brief History of Leveling at the National Geodetic Survey"[45] and the links provided there.

Prior to our understanding of the constraints, elevations

were certainly surveyed, but only relatively—surveyors knew that point A was higher than point B and by how much, but there wasn't a standard datum to connect one surveyed area with another. It wasn't until the mid to late 1800s that instrumentation technology had advanced to enable elevation measurements across the continent and along railroads and rivers tied to tide gauges. Eventually, the "point A to point B" surveys and continental surveys were tied together and connected to a standard reference point for mean sea level as the datum. This network has to be periodically adjusted to account for changes in the land surface due to postglacial uplift, earthquakes, and subsidence. Recent modernization has included the integration of satellite GPS technology with the network for quick and accurate elevation measurements.

As mentioned above, the land surface is not stable and fluctuates in elevation with time due to several factors. Sea level is certainly not a stable datum either given the variability caused by tidal and weather influences and the fact that by definition, it is present only due to its juxtaposition with the land. The reason sea level is used as a datum is because between the two (land and sea), sea level is believed to be the more stable due to the properties of water,[46] the main one being that water seeks its own level and forms a flat surface when not acted upon by other forces. It is for this "very real" reason that climate change alarmists fear melting of ice at the poles and an associated rise of water along coastlines.

However, it has always been difficult to discern whether it is the land or the sea that is moving. Sea levels may appear to be rising when actually the land surface is sinking and vice versa. In fact, everything is moving. There is no part of Earth, our solar system, our galaxy, or the universe that is stable. That is why even the integration of satellite GPS into the methodologies for surveying elevations makes the elevations only more relatively accurate; satellites are not perfectly stationary.

Sea level may very well be rising, but the whole surface of Earth moves up and down and around due to tectonic, tidal, and buoyancy effects. Add to that the relative nature of elevation surveys and there is a lot to be learned before we can definitively state that sea level is rising and much more that it's due to climate variability.

Extinction

A strong correlation exists between historical CO_2 levels in the atmosphere and the timing of extinction events.[47] All life and other organic processes contribute to CO_2 in the atmosphere. Either increases in the abundance of life between the ice ages led to increasing CO_2 levels and then to climate variability and extinction events and correspondingly further increases in CO_2 levels, or astronomical perturbations led to global warming, causing extinction events, which lead to increases in CO_2 levels. There are other scenarios, but that's one of the reasons the concept of man-made climate variability is questioned.

One graph shows the percentage of marine animal genera going extinct during any given time span.[47] Note that extinctions corresponding to the Permian-Triassic and

Triassic-Jurassic boundaries appear to match exactly to increases in CO_2 levels in the same time intervals;[30] the graphs over all are spookily concomitant. Did warming caused by the proliferation of animal life lead to increasing CO_2 and ultimately to global extinction? Again, this is just one possible scenario.

If that were the case, CO_2 levels may have spiked to higher levels than those shown similar to the kind of spike we may be seeing now;[48] we can't measure the higher levels that would have occurred during the extinctions because data in the geologic record used to estimate CO_2 levels have degraded naturally. They degrade due to chemical reactions in the soil and water to a lower state due to the natural orders: for example, a rock weathers (or erodes) down to its sediments.

Note also that the intensity of extinctions decreases with Earth's age. This indicates that life becomes less diverse with the increasing age of Earth. Whether the extinctions are caused by climate variability or vice versa, the combination of events appears to occur cyclically. Should we try to minimize it?

Sure. It seems there are two extremes to our beliefs about climate variability—that it's not occurring or that it's really bad and entirely our fault. We should be somewhere in the middle. Climate variability may be a cycle of life in the larger scheme of things that only stimulates our understanding of our ability to control it. Turns out nature always manages to have a life of its own.

Footprints

Our (human) carbon footprint is determined by our conveniences. The conveniences we enjoy are determined by our status in society and our achieved standards of living. There is a limit to how much of our accomplishments we want to give up; we do not want to make our own butter if we don't have to.

Take for example the sought-after transition to solar energy, which is a result of our industriousness. We have come up with a way to collect energy from the Sun and produce electricity. How convenient! But the manufacture of solar panels comes with its own carbon footprint as do other forms of renewable energy production.[49] We use the energy and components of fossil fuels to make solar panels. If we didn't want the convenience of electricity, there would be no need to create this carbon footprint or any additional carbon footprint at all. With increasing technology, it is increasingly possible to decrease the carbon footprint due to production over the life of the product; same however with the production of fossil fuels.

Who and what decides where we draw the line? The use of coal to generate electricity does have a larger carbon footprint compared to the equivalent in solar panels. However, carbon-capture technology could bring the coal footprint down to within the same order of magnitude as solar's. Coal has other advantages, however; it's plentiful and is a resource that shouldn't be ignored.

We will continue to use electricity to charge our batteries we develop to replace the use of fossil fuels, but batteries also have a carbon footprint. To completely replace coal and the use of fossil fuels, all the things we manufacture to produce electricity in their place also have carbon footprints.

However, the resources we utilize (oil, gas, and coal) have their own conveniences in that they are plentiful and necessarily utilized for the value they provide. Seas of coal and oil are useful for no other purpose; they were put here to be exploited.

The Gases

What came first; oxygen or carbon dioxide? The world started out with more carbon dioxide and other gases and very little oxygen; oxygen didn't come until plants gained a foothold and started producing it. Then for the most part, the animals came and started consuming the oxygen and producing carbon dioxide.

You may believe from the headlines that humans produce all of the CO_2 by burning fossil fuels, but all animals down to the smallest ant produce CO_2; that's a fact of nature. Plants also release CO_2 in addition to the oxygen they produce. By far, the largest source of CO_2 along with terrestrial (soil and vegetation), residual atmospheric CO_2, and volcanic activity are the oceans. Conservatively, these natural sources of CO_2 combined account for approximately 97 percent of the CO_2 available.

Carbon dioxide is released and absorbed by the soil, vegetation, and oceans; it is not easily held in place. With warming, more CO_2 is lost from the natural sources and less is absorbed. To be fair, supposedly, the amount of CO_2 released by the oceans and terrestrial systems is in balance with what is absorbed; the small amount (approximately 3 percent) produced by burning fossil fuels accumulates (is not reabsorbed), again supposedly, thereby leading to the additional warming.

This begs the question: what came first, CO_2 or warming? If Earth has been warming since the last glacial period as seen in the geologic record, it stands to reason that CO_2 levels would also be increasing; that is the natural cycle.

We don't know what the ozone hole was doing more than 40 or 50 years ago, much less during periods like the "dust bowl," or during previous warming periods over the Earth's long history. The Earth started out with no ozone or oxygen, (and more carbon dioxide) but gradually developed to the state where plants could evolve, which started spewing oxygen into the atmosphere. Animals evolved and started using oxygen and spewing carbon dioxide.

Thus began a vicious cycle—the biomass of oxygen-consuming (and carbon dioxide–creating) species increasing beyond the biomass of carbon-dioxide consuming (and oxygen-creating) species. It is difficult to say whether the planet's warming and cooling cycles are related to the biomass cycle or vice versa. But undoubtedly, during the warming periods, the amount of ozone in the atmosphere decreases due to the decrease in the oxygen/carbon dioxide ratio.

Matrimony

Is global warming such a bad thing? To our spouses who say they are "too cold" more often than they say they are "too warm," the concept of global warming is undoubtedly a godsend. Sure, they complain when it's hot but not as much as when it's cold.

This concept is not lost on the world. Freezing temperatures are more uncomfortable and more inconvenient than are warming temperatures: it's much easier to shed clothes than to pile them on. No one vacations on Alaskan beaches except maybe someone who lives farther north. Ask a homeless person where he would rather sleep—Florida or Wisconsin.

Do the global warming alarmists have an agenda that goes above and beyond, or should I say underneath, the benefits of a warming planet? Do they have something to gain by convincing the world (and our spouses) that it is better to have a planet that cools to a greater extreme than it warms?

Freezing temperatures are also more deadly than are warming temperatures.[50] There are up to fifteen times more deaths from freezing temperatures than there are from warming temperatures. Life proliferated during the last warming trend until it was wiped out by that comet. Carbon dioxide was in great supply *and* in great demand.

Of course, there are other disadvantages to a warming planet; a rise in sea level gives our governments the most grief. But coastal and floodplain communities have been dealing with flooding for a long time. There are no better ways to deal with it than to avoid trying to live there in the first place or to relocate to higher ground.

So is global warming a bad thing? The dire predictions of impending increases in severe weather have no basis. The weather may actually improve, or it may shift from one region to another. Either way, there will be decreases in the death rate from freezing and less complaining from our mates.

We may eventually understand our planet, its workings, and our relationships with the interactions that sustain and challenge the living. We may even eventually understand our solar system and our relationship to it. We may never understand our galaxy much less the universe that extends beyond. We can only surmise and imagine from our existence outward.

As far in space as we can see and as long in time as we know, we are the dominant species in the universe, beside the Creator. We have a destiny and obligation to manage the energy from creation to our advantage and advancement.

References

1. http://phys.org/news/2013-04-sunlight-earth.html.
2. https://en.wikipedia.org/wiki/Earth.
3. https://en.wikipedia.org/wiki/Observable_universe.
4. http://www.universetoday.com/65644/how-far-is-a-lightyear-in-miles/.
5. https://en.wikipedia.org/wiki/Alpha_Centauri.
6. http://www.universetoday.com/65601/where-is-earth-in-the-milky-way/.
7. http://imagine.gsfc.nasa.gov/features/cosmic/nearest_galaxy_info.html.
8. http://imagine.gsfc.nasa.gov/features/cosmic/milkyway_info.html.
9. http://www.universetoday.com/26623/how-fast-does-the-earth-rotate/.
10. https://en.wikipedia.org/wiki/Abundance_of_the_chemical_elements.
11. http://www.scholastic.com/teachers/article/solar-system-0.
12. https://www.ncdc.noaa.gov/monitoring-references/faq/anomalies.php.

13. https://en.m.wikipedia.org/wiki/Foot_(unit).
14. http://mathcentral.uregina.ca/QQ/database/qq.09.96/kredo1.html.
15. https://en.m.wikipedia.org/wiki/Inch.
16. http://www.currentresults.com/Environment-Facts/changes-in-earth-temperature.php.
17. https://en.m.wikipedia.org/wiki/Lowest_temperature_recorded_on_Earth.
18. http://www.answers.com/mobile/Q/What_is_the_hottest_air_temperature_ever_recorded_on_earth.
19. https://en.m.wikipedia.org/wiki/List_of_climate_scientists.
20. https://en.m.wikipedia.org/wiki/List_of_scientists_opposing_the_mainstream_scientific_assessment_of_global_warming.
21. http://www.earth-climate.com/.
22. http://climate.nasa.gov/evidence/.
23. http://cdiac.ornl.gov/trends/co2/ice_core_co2.html.
24. http://www.geocraft.com/WVFossils/greenhouse_data.html.
25. http://ar5-syr.ipcc.ch/ipcc/ipcc/resources/pdf/IPCC_SynthesisReport.pdf.
26. https://en.m.wikipedia.org/wiki/Little_Ice_Age.
27. https://en.m.wikipedia.org/wiki/Weather_forecasting.
28. https://disc.gsfc.nasa.gov/ozone/keep-for-review/pre-codi/ozone_measurements.html.
29. https://friendsofscience.org/assets/documents/FoS percent20 Pre-industrial percent20CO2.pdf.
30. https://en.wikipedia.org/wiki/Ice_age.

31. https://en.wikipedia.org/wiki/Temperature_record_of_the_past_1000_years.
32. https://www.ncdc.noaa.gov/paleo/globalwarming/paleolast.html.
33. http://www.climate-skeptic.com/temperature_measurement/.
34. https://en.m.wikipedia.org/wiki/Ice_core.
35. http://www.ncdc.noaa.gov/paleo/reports/trieste2008/ice-cores.pdf.
36. https://en.m.wikipedia.org/wiki/Geologic_temperature_record.
37. https://en.m.wikipedia.org/wiki/Dendroclimatology.
38. http://www.climate-skeptic.com/temperature_history/.
39. http://www.skepticalscience.com/Tree-ring-proxies-divergence-problem.htm.
40. http://www.climatecentral.org/news/five-year-forecast-more-warming-in-store-19988?utm_content=buffer485d5&utm_medium=social&utm_source=twitter.com&utm_campaign=buffer.
41. http://www.srh.noaa.gov/jetstream/global/jet.htm.
42. https://ams.confex.com/ams/pdfpapers/134671.pdf.
43. http://m.livescience.com/44881-jet-stream-history-north-america.html.
44. https://en.m.wikipedia.org/wiki/Jet_stream.
45. http://celebrating200years.noaa.gov/foundations/leveling/welcome.html#intro.
46. http://water.usgs.gov/edu/water-facts.html.
47. https://en.wikipedia.org/wiki/Extinction_event.

48. http://climate.nasa.gov/evidence/.
49. http://spectrum.ieee.org/green-tech/solar/solar-energy-isnt-always-as-green-as-you-think.
50. http://sharpgary.org/Warm percent20Vs percent20Cold percent20Deaths.html.

Appendix

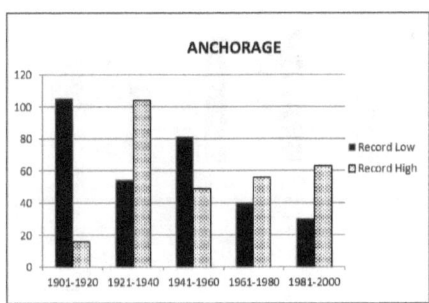

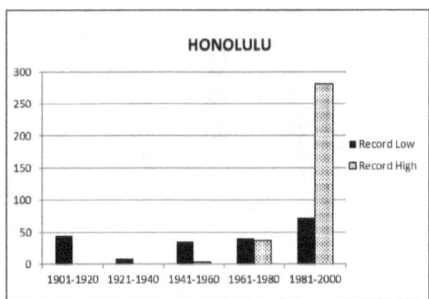

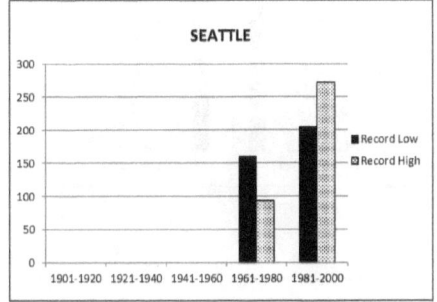

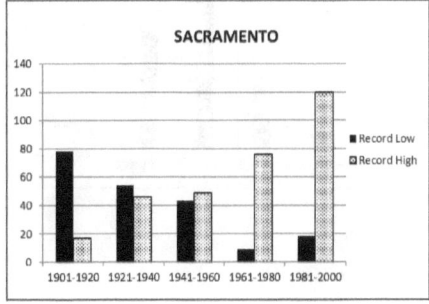

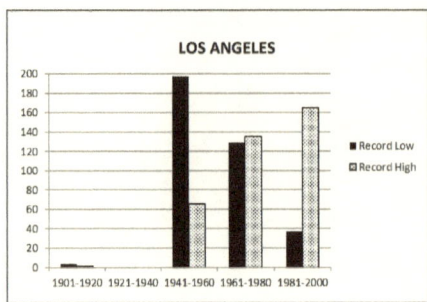
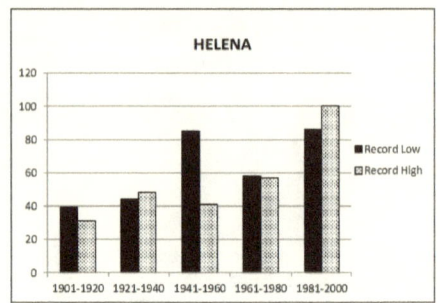
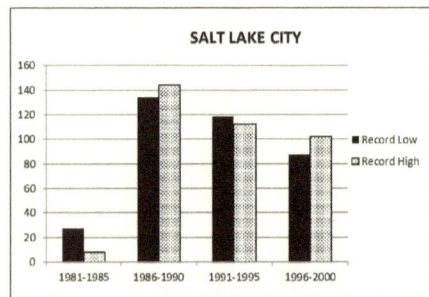
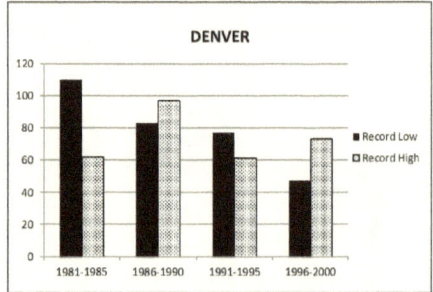
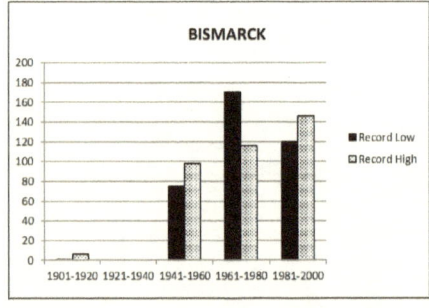
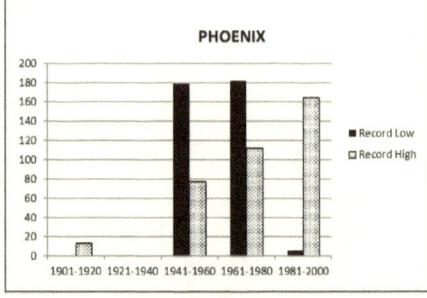
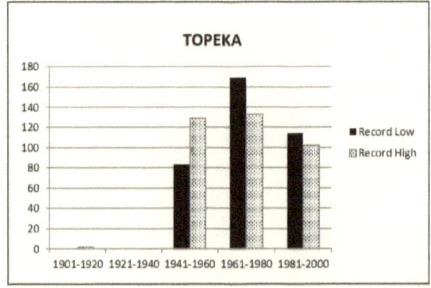
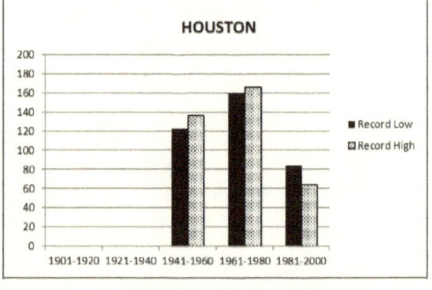

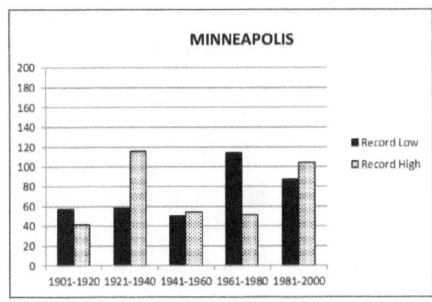
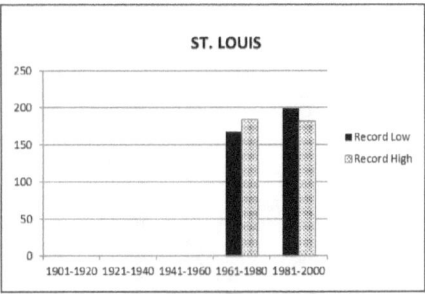
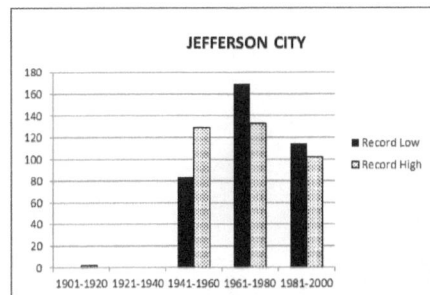
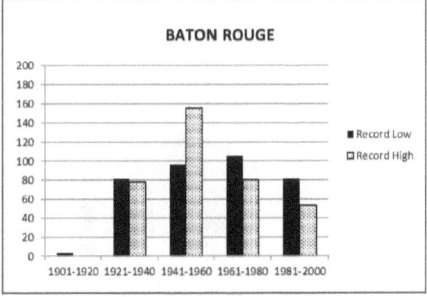
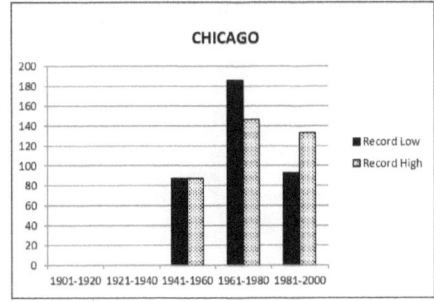
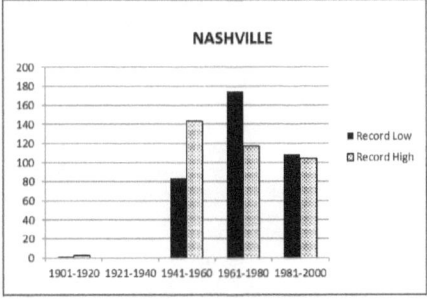
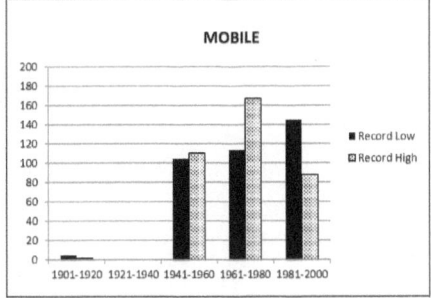
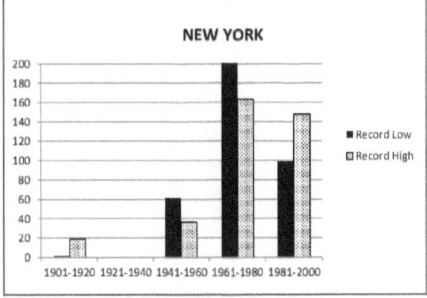

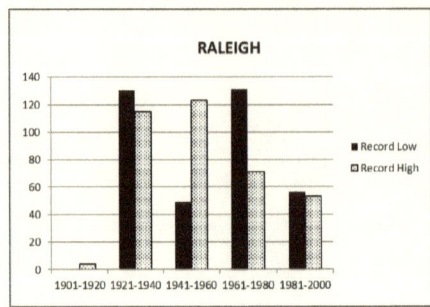
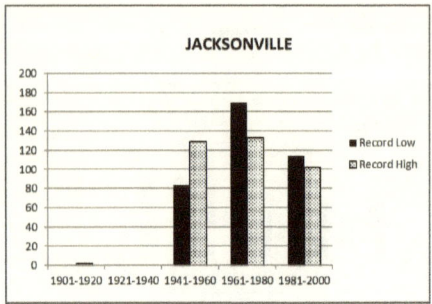
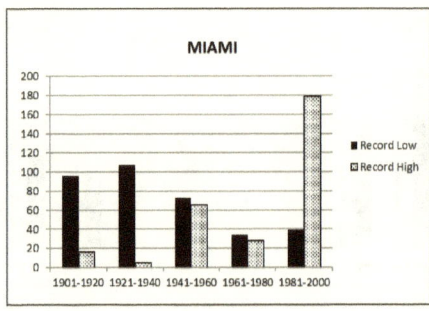

www.ingramcontent.com/pod-product-compliance
Lightning Source LLC
Chambersburg PA
CBHW030857180526
45163CB00004B/1619